W9-BIY-355

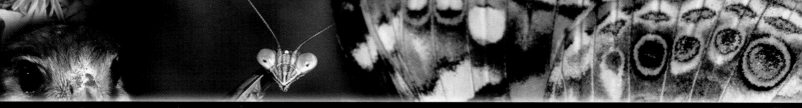

THE LIBRARY OF FOOD CHAINS AND FOOD WEBS™

Food Chains in a
MEADOW HABITAT

ISAAC NADEAU
Photographs by
DWIGHT KUHN

The Rosen Publishing Group's
PowerKids Press™
New York

To Nellie: meadow walker, language lover, grandma — Isaac Nadeau
To Meghan's friend Zoe — Dwight Kuhn

Published in 2002 by The Rosen Publishing Group, Inc.
29 East 21st Street, New York, NY 10010

First Edition

Book Design: Emily Muschinske
Project Editor: Emily Raabe

All photos by Dwight Kuhn.

Nadeau, Isaac.
Food chains in a meadow habitat / Isaac Nadeau.
    p. cm. — (The Library of food chains and food webs)
Includes bibliographical references (p.   ).
  ISBN 0–8239–5762–4 (lib. bdg.)
1. Meadow ecology—Juvenile literature. 2. Food chains (Ecology)—Juvenile literature. [1. Meadow ecology.
2. Food chains (Ecology) 3. Meadows. 4. Ecology.] I. Title. II. Series.
  QH541.5.M4 N235 2002
  577.4'616—dc21
                                                                    2001000618

Manufactured in the United States of America

Contents

Food Chains Are Everywhere!

Do you know what a food chain is? A food chain is a way to see how living things are connected by what they eat. To live, plants and animals need energy. A dandelion growing in a meadow gets its energy from the Sun. A grasshopper eats the dandelion. A praying mantis eats the grasshopper for energy. The praying mantis may be eaten by a hawk. When the hawk dies, there are thousands of insects and **bacteria** in the soil waiting to gobble it up. The dandelion, grasshopper, praying mantis, hawk, insects, and soil bacteria are all links in the food chains of a meadow.

Each time the energy from the Sun is passed on, another link is added to the food chain. If you visit a meadow, you can see food chains all around you. In fact, every living thing you see in the meadow is part of a food chain.

This dandelion, grasshopper, praying mantis, and sparrow hawk make up a food chain.

What Is a Meadow?

A meadow is a field that is covered with grasses and other plants without many trees on it. Of all the things that live in a meadow, one of the first things that you notice are the grasses. There are many different kinds of grasses in a meadow. Some are no taller than your knees. Others are taller than many adults. Some meadows are high up in the mountains. Many meadows can be found where a farmer once cleared the land to plant his crops, or where cows or sheep were let loose to graze. The meadow is home to all kinds of interesting animals, too. These animals are able to find everything they need in the meadow, including food, shelter, and water. A place that has everything a plant or an animal needs to live is called a habitat. The meadow is a great habitat for many animals and plants.

Meadows are covered in grasses. There are few, if any, trees in a meadow habitat.

Working for a Living

When a butterfly visits a flower in a meadow, it is looking for a sugary juice called nectar. When the butterfly drinks the nectar, it gets pollen on its feet or head. The next time the butterfly visits a flower, some of this pollen is rubbed off onto the new flower. With this pollen, a seed can begin to ripen inside the new flower.

Have you ever wondered why the flowers in a meadow smell so good and have such beautiful colors? These flowers are trying to show off, but not for people. The flowers are trying to attract the insects and birds that help to spread their pollen. Pollen grains are the tiny specks that flowers use to **reproduce**. When pollen grains from one flower land on another flower of the same **species**, the seeds inside that flower begin to ripen. This is called **pollination**. Pollination is very important in the meadow food chain. Without it, new flowers could not grow in the meadow, and many insects would go hungry. All over the meadow, flowers and insects depend on each other.

8

Bees, butterflies, moths, hummingbirds, flies, beetles, and ants are all animals that help flowers through pollination.

Setting the Table

Early in the morning, as soon as the Sun comes out, all of the plants in the meadow begin to do their work. Inside the leaves of velvet grass, oxeye daisy, and all the other meadow plants, an amazing event takes place. The plants turn sunlight into food! This important work is called **photosynthesis**, and only plants can do it. When sunlight strikes the leaves, it mixes with water and air to produce sugar. Plants use the energy from this sugar to grow new leaves, stretch their roots, and make flowers and **nectar**. Plants are called **producers** because they produce all the energy for the food chains in the meadow. Plants are the first link in a meadow food chain. If there were no plants, there would be nothing for the hungry groundhogs, grasshoppers, and other meadow **herbivores** to eat.

Producers such as these meadow grasses use the energy from the Sun to make food.

Plants on the Meadow Menu

For every plant in the meadow habitat, there is an herbivore that likes to eat it. Herbivores are the middle link in the meadow food chain, between the plants they like to eat and the animals that like to eat them. The meadow vole, for example, is a meadow herbivore who eats grass, clover, roots, and other plants. It eats almost its entire weight in plants every day! For every plant the vole eats, however, there is a **predator** that likes to eat voles. Meadow voles have to watch out for foxes, coyotes, hawks, owls, cats, and dogs. These animals eat so many voles that the voles must have lots of babies each year just to keep up. A single meadow vole can have as many as 30 babies in a single year, but only a few of these will survive long enough to have their own babies.

When a white-tailed deer fawn (inset), a woodchuck (inset) or meadow vole (large picture) munches on a plant, the energy stored in the plant is passed into the deer, the woodchuck, or the vole.

Meadow Carnivores

The ambush bug is a meadow carnivore. It gets its name because it hides inside flowers waiting for a bumblebee or hornet to land. The ambush bug leaps out and attacks its prey with its claws, then sucks the juice out of it.

Some of the hardest workers in the meadow are the predators. Predators have to catch their food. The animals that they catch are called **prey**. It is not always easy to catch your food, especially when it has feet or wings! Predators that eat other animals are called **carnivores**. Carnivores are the third link in a meadow food chain.

Foxes and coyotes are two carnivores that find plenty of food in the meadow habitat. They pounce on voles, shrews, mice, and other small meadow animals and hold them between their front paws. If foxes get very hungry, they will eat insects or even plants.

Foxes have such good hearing that they can hear a mole digging its tunnel underground!

No Picky Eaters Here

There is a lot to eat in a meadow food chain. For some animals, it must seem like everywhere they turn, they're looking at another meal. Animals that eat many different foods, including plants and other animals, are called **omnivores**. Humans are a good example of omnivores that might visit a meadow habitat. The opossum is also an omnivore. The opossum eats berries, grubs, insects, eggs, and small animals. Some omnivores eat the bodies of dead animals. These omnivores are called **scavengers**. Many scavenger insects lay their eggs in the bodies of dead animals. When the eggs hatch, the **larvae** feed on the animal. Scavengers help break down large animal bodies into smaller parts. When a plant or an animal dies, it will not be long before it becomes part of the soil again.

This burying beetle will bury this dead shrew in the dirt, and when its young hatches underground, they will eat the buried shrew.

Not Just Dirt

Almost everything in the meadow habitat is food. You can even find an animal that likes to eat animal droppings. This animal, called a tumblebug or dung beetle, is part of a very important group of living things called **decomposers**. Most decomposers live in the soil. Decomposers return many important **nutrients** to the soil. These nutrients help plants grow. Tiny animals called springtails are examples of decomposers. Springtails live in the soil and take bites out of dead plants. Tiny worms also break down the dead plants that have fallen to the ground. Finally bacteria and **fungi** finish the job. All of these decomposers are part of the food chains in the meadow habitat. They give nutrients back to the soil, but they also are eaten by other animals, such as beetles and other insects, that live in or near the soil.

This dead fly is being covered in mold. The mold will help to decompose the fly's body back into the soil.

Making the Connection

All through the meadow and all over the world, plants, animals, and other living things depend on each other to stay alive. Herbivores need plants, carnivores need herbivores, scavengers need dead animals, decomposers need dead plants and animals, and plants need decomposers. Every link in the food chain depends on each of the others. The meadow can teach us that every living thing has an important place in the world.

Color Key

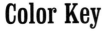 carnivores

decomposers

herbivores

omnivores

producers

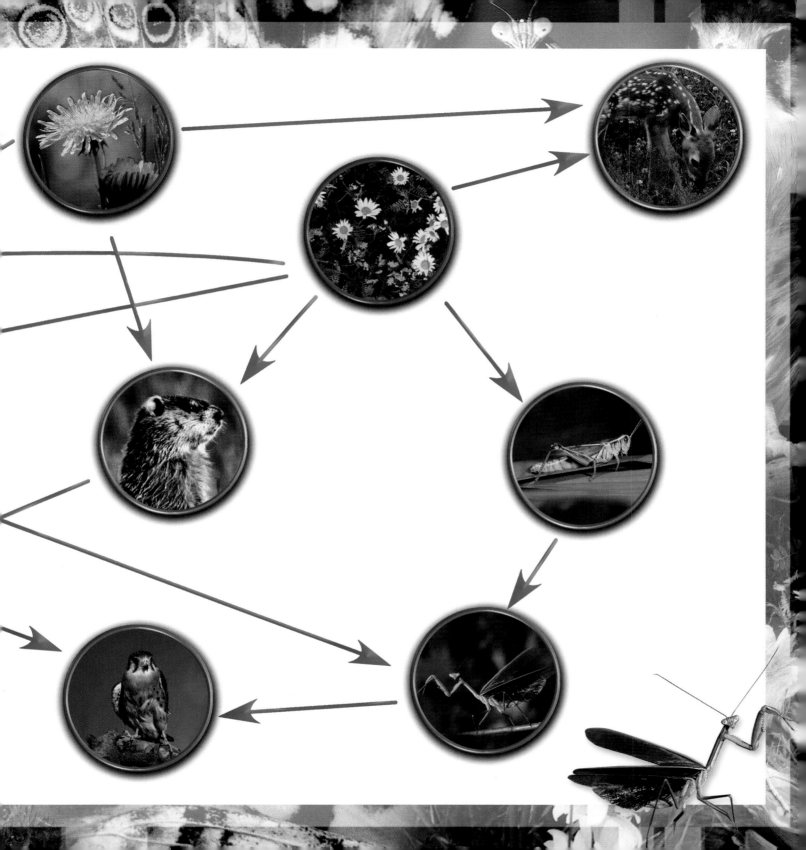

A Trip to the Meadow

A naturalist is a person who goes outside to pay careful attention to the natural world. The meadow habitat is a good place to go to learn to be a naturalist. There are a great many meadows, and each one has a lot of different living things to see.

If you are lucky, you might get a chance to visit a meadow. You might even live near one. There are many questions you might ask yourself as you explore the meadow. For example, how does the meadow change from winter to summer? How many different kinds of insects can you find? On what day does the first flower bloom in the spring? Do you see different animals at different times of day? Can you find a producer, an herbivore, a carnivore, an omnivore, and a decomposer in the meadow? If you cannot go to a meadow, look in a book or on the Internet. There are many more amazing facts that you can learn about the meadow habitat.

Glossary

bacteria (bak-TEER-ee-uh) Tiny living things that can only be seen with a microscope and that cause living things to decay.

carnivores (KAR-nih-vorz) Animals that eat other animals for food.

decomposers (dee-kum-POH-zerz) Organisms, such as fungi, that break down the bodies of dead plants and animals.

fungi (FUN-jeye) A group of living things that feed off waste and dead things. Fungi are like plants, but they do not photosynthesize.

herbivores (ER-bih-vorz) Animals that eat plants for food.

larvae (LAHR-vee) Baby insects.

nectar (NEK-tur) A sweet, sugary liquid made by flowers to attract insects and birds.

nutrients (NOO-tree-ints) Anything that a living thing needs to live and grow.

omnivores (AHM-nih-vorz) Animals that eat both plants and animals for food.

photosynthesis (foh-toh-SIN-thuh-sis) The process in which leaves use energy from sunlight, gases from air, and water to make food and release oxygen.

pollination (pah-lih-NAY-shun) When a flower gets another flower of the same kind's pollen on it, and a seed begins to ripen.

predator (PREH-duh-ter) An animal that hunts other animals for food.

prey (PRAY) An animal that is hunted by another animal for food.

producers (pruh-DOO-serz) Plants that use sunlight to make energy available for use by living things.

reproduce (ree-pruh-DOOS) To make more of something of the same kind.

scavengers (SKA-ven-jurz) Animals that feed on dead animals.

species (SPEE-sheez) A group of plants, animals, or other living things that can only reproduce with each other, and not with members of another species.

Index

B
bacteria, 4, 18

C
carnivore(s), 14, 22

D
decomposer(s), 18, 22

E
energy, 4, 10

F
fungi, 18

G
grass(es), 6, 10, 12

H
habitat, 6
herbivore(s), 10, 12, 22

L
larvae, 16

N
nectar, 10

O
omnivore(s), 16, 22

P
photosynthesis, 10
pollen, 8
pollination, 8
predator(s), 12, 14

prey, 14
producer(s), 10, 22

R
reproduce, 8

S
scavengers, 16
soil, 4, 16, 18
species, 8
Sun, 4, 10

Web Sites

Due to the changing nature of Internet links, PowerKids Press has developed an online list of Web sites related to the subject of this book. This site is updated regularly. Please use this link to access the list:

www.powerkidslinks.com/lfcfw/meadow/